Herbs for Healthy Digestion

Karen Bradstreet, M.H.

Woodland Publishing
Pleasant Grove, UT

©1997
Woodland Publishing
P.O. Box 160
Pleasant Grove, UT 84062

The information in this book is for educational purposes only and is not recommended as a means of diagnosing or treating an illness. All matters concerning physical and mental health should be supervised by a health practitioner knowledgeable in treating that particular condition. Neither the publisher nor author directly or indirectly dispense medical advice, nor do they prescribe any remedies or assume any responsibility for those who choose to treat themselves.

Contents

The Importance of Healthy Digestion	5
How the Digestive Process Works	6
Digestive Organs	7
Enzymes	8
Common Causes of Digestive Problems	
Eating the Wrong Types of Foods	9
Stress	10
Lactose-Intolerance	10
Not Chewing Food Well	11
Food Allergies	11
Enzyme Deficiency	12
Overeating and Constant Snacking	13
Imbalance of Beneficial Bacteria	13
Poor Elimination	14
Natural Remedies for Common Digestive Problems	
Constipation	15
Diarrhea	18
Flatulence	20
Heartburn	22
Ulcers	23
Irritable Bowel Syndrome (IBS)	26
Diverticulosis	28
Hemorrhoids	30
Conclusion	31

Introduction

Eat. Digest. Eliminate. That's the digestive process in a nutshell. But what should be a simple bodily function is often a source of distress for almost everyone at some time in their life. Even worse, for many it can also be the source of debilitating illness. Those suffering from digestive problems spend millions of dollars each year on antacids, ulcer medications, laxatives, antidiarrhea preparations and other drugs. Unfortunately, most drugs simply mask symptoms and don't resolve the underlying problems.

Armed with the information found in these pages, you can solve digestive problems where they start. This booklet contains easy-to-use suggestions, dietary tips, herbal remedies and other supplements you can use to enjoy a lifetime of healthy digestion—naturally.

The Importance of Healthy Digestion

The digestive process is fundamental to life. In simplest terms, the digestive system's job is 1) to take in food, 2) to break it down so nutrients can be absorbed as a source of life-giving energy, and 3) eliminate the waste. The health of the digestive system has far-reaching effects on overall health. If nutrients are not properly broken down and absorbed, the body can't adequately fuel or repair itself. Everything suffers, from the smallest cells to the largest organs and tissues.

Energy production and nutrient absorption aren't the only important functions of the digestive system. Proper elimination is also essential to health and vitality. If uneliminated waste is allowed to stagnate in the colon, the body can become poisoned. Many health problems begin in the colon. If problems in the colon aren't corrected, toxins spread to other tissues and organs, resulting in a wide variety of diseases. The list of potential digestive problems is a long one that includes indigestion, nausea, vomiting, hiatal hernias, peptic and duodenal ulcers, irritable bowel syndrome, ulcerative colitis, diverticulitis, constipation, diarrhea, and hemorrhoids. Because most digestive problems are linked to poor nutrition, they respond exceptionally well to proper nutrition and supplementation.

How the Digestive System Works

Every day, the average adult processes about two-and-a-half gallons of food through some 30 feet of digestive organs. The digestive system consists of the mouth, the esophagus, the stomach, the small and large intestines, and the rectum. The liver, gall bladder and pancreas also play essential roles. Enzymes, too, have an integral function in proper digestion. Some enzymes are produced by the body, while others are present in food. Digestive problems can occur in any of these areas.

The Mouth and Esophagus

Located in the mouth are three pairs of salivary glands which are stimulated by chewing. As food is mixed with saliva, it is lubricated for easier swallowing and begins to dissolve in the first step of digestion. An enzyme in saliva, called ptyalin, converts complex carbohydrates into simple sugars. Simple sugars are further broken down into glucose, which provides energy for all body processes. It is essential to chew food thoroughly to give the enzymes in saliva a chance to permeate food. Poorly chewed food is a prime cause of bloating and stomach discomfort after eating. Chewing also stimulates the stomach, intestines and other digestive organs, preparing them for their function in the digestive process.

Herbs for Healthy Digestion

After food is swallowed, it travels down the throat to the esophagus, a muscular tube that pushes food along to the stomach. The esophagus is the most common site of heartburn.

The Stomach

In the stomach, food is churned and mixed into a paste called chyme to allow for further digestion. Digestive juices in a strongly acidic solution—most notably, hydrochloric acid (HCl)—further break down food. Hydrochloric acid is so strong it can dissolve a razor blade, but in a healthy stomach mucus secretions keep it from damaging the stomach lining.

The stomach wall consists of three layers of muscles which move in rhythmic contractions to mix food with digestive juices and then send it along in the digestive process. The stomach is where protein digestion begins and where some food components, including sugars, are directly absorbed into the bloodstream to provide fuel for energy.

From the stomach food passes into the duodenum, where digestion continues with the help of secretions from the gallbladder. Next, food passes into the small intestine.

The Small Intestine, Liver, Gallbladder & Pancreas

In the small intestine digestion of starches, proteins and fats occurs with secretions from the liver, gallbladder, pancreas and intestines. About 90 percent of nutrient absorption takes place in the small intestine.

Each day, the liver manufactures anywhere from a pint to a quart of a bitter alkaline substance known as bile, which is stored in the gallbladder. Bile is released into the small intestine and, with the help of enzymes, breaks down fatty materials.

The pancreas is located behind the stomach in the upper left side of the abdomen. This important organ secretes enzymes and hormones—one of the most well-known is insulin—which are needed for the digestion and absorption of food.

The Large Intestine and Rectum

Components of food that are not absorbed in the small intestine are moved into the large intestine. More nutrients and water are absorbed and then remaining waste matter is eliminated through the rectum.

Generous amounts of dietary fiber are essential for colon health. Fiber helps speed the passage of waste through the colon and out of the body for elimination. When fiber-poor foods are standard fare in a person's diet, uneliminated food can build up and stagnate, creating numerous problems for the colon and, eventually, the whole body. Low fiber consumption is linked to almost all diseases of the colon, as well as many types of cancers.

A healthy intestine is home to millions of beneficial bacteria that assist in the absorption of nutrients and manufacture vitamins, including vitamin K and the B-complex family. They also help keep harmful bacteria in check. The most well-known of these friendly bacteria is *Lactobacillus acidophilus,* present in some fermented foods such as yogurt. Antibiotics kill beneficial bacteria along with harmless ones, allowing yeasts and other harmful organisms to proliferate. This can increase susceptibility to food allergies, digestive discomforts, and infection from harmful microorganisms.

Enzymes

Enzymes are remarkable protein substances that act as catalysts in countless actions in the body, including the digestion of food and repair of tissue. Each of the thousands of enzymes in the body has a specialized function. Without enzymes, life could not exist. The body manufactures some enzymes, but it also gets enzymes from food. Even low levels of heat destroy food enzymes, making it important to eat plenty of raw foods. If one eats a diet devoid of raw foods, the body's supply of enzymes becomes overtaxed, increasing susceptibility to low energy, illness and disease. Enzyme supplements are available through many manufacturers. Look for a product that contains the following key enzymes, plus hydrochloric acid and bile salts:

Pepsin: Digests proteins.
Pancreatin: Produced by the pancreas; digests proteins, carbohydrates, and fats.
Mycozyme: Digests starches.
Papain and Bromelain: Digest protein.
Lipase: Digests fats.

Summary of the Digestive System

Because the digestive system supplies fuel for all bodily functions, it is essential to eat foods and use supplements that build, rather than hinder, the digestive process. Most problems with the digestive system can be traced to poor dietary choices. The next section discusses common causes of digestive problems in more detail.

COMMON CAUSES OF DIGESTIVE PROBLEMS

The first step in solving digestive problems is to determine where they begin. Listed here are the most common causes of problems, with suggested solutions. Follow these general guidelines if you suffer from digestive problems. Then add appropriate herbs and supplements (listed in the next section) to target specific problems or weaknesses.

Eating the Wrong Types of Food

Heavily processed foods lack the enzymes necessary for digestion and the fiber necessary for healthy elimination. The average American diet is woefully lacking in fresh foods. A five-year study of American dietary habits revealed that on an average day, 45 percent of Americans eat no fruit or juice, 48 percent eat no vegetables, and only 9 percent eat the recommended minimum amount of fruits and vegetables daily.[1] That's 91 percent of Americans who aren't even getting the minimum amount of fresh foods the body needs.

As for fiber, people whose diets are low in fiber are especially susceptible to problems with the lower bowel. Colon and rectal cancer are the second leading cause of death in the United States. In Africa, the average diet has seven times the fiber of the average American diet, and colon/rectal problems are rare there. In addition to preventing colon/rectal cancer, fiber helps prevent constipation, diverticulosis, and most other illnesses of the lower bowel. In order to avoid these problems, some simple solutions are:

- Eat fresh, raw fruits and vegetables as much as possible, in addition to whole grains and legumes.
- Eat whole wheat bread. Some bread labeled whole wheat is simply white flour tinted brown by using molasses as a sweetener. Be sure to read the ingredients label on the bread you buy. Look for "100% stone ground wheat" rather than enriched flour.
- Take enzyme supplements with meals to help your body break down and assimilate food, especially processed food.
- Take a daily fiber supplement (those containing psyllium hulls are an excellent choice).

Stress

Stress has profound effects upon the body in general, and the digestive process in particular. It affects digestive secretions and the contractions of the stomach and intestines. Problems related to eating while under emotional stress range from stomach discomfort to irritable bowel. A simple solution to this problem is to not eat when excessively stressed, or if you must eat, eat lightly.

Lactose Intolerance

Those who are lactose-intolerant have difficulty digesting milk products because they do not produce enough of the enzyme lactase, which breaks down lactose (milk sugar) in the intestines. Such individuals suffer from gas, stomach cramps and diarrhea when they consume milk products.

Herbs for Healthy Digestion

Starting in early childhood, most people start producing less lactase. Only about 30 percent of all people maintain the ability to digest lactose through adulthood. People most likely to be lactose-intolerant are Asians, those from West Africa, about 80 percent of Native Americans, 50 percent of Hispanics, and 20 percent of Caucasians.

You can test yourself to determine whether you have digestive discomfort caused by lactose-intolerance. Drink a couple glasses of milk on an empty stomach, and see if symptoms occur within the next three or four hours. If they do, try the experiment again with lactase-treated milk. If symptoms don't occur the second time, you're probably lactose-intolerant. Some simple solutions to the problem of lactose intolerance are:

- Buy milk products designed for the lactose-intolerant
- Try fermented milk products such as yogurt, which contain less lactose.
- Use a lactase supplement when you consume milk products. You can purchase liquid and add it to milk products before eating, or take tablets before you consume a milk product.

Not Chewing Food Well

Failure to chew food well results in the food not being permeated by enzymes in saliva, setting the stage for gurgling and bloating when food reaches the stomach. Not chewing well is a common cause of burping, bloating, and gas. A simple solution to these problems is to avoid eating in a rush. Take time to enjoy the flavor of your food. Chew slowly.

Food Allergies

Food allergies occur when the body produces antibodies, most notably the antibody IgE, that fight against ingested foods. This antibody, found in the intestinal wall, causes gas, bloating, and severe pain. Some foods produce an immediate reaction, while oth-

Herbs for Healthy Digestion

ers cause a delayed reaction. Foods that produce an immediate reaction are easy to identify and eliminate, but more subtle allergies are harder to detect.

There are a wide variety of symptoms of food allergy not limited to the intestinal system, including acne, arthritis, asthma, depression, fatigue, headaches, insomnia, weight gain, and sinus problems. Colitis, intestinal discomfort, and ulcers are symptoms that strike the digestive system.

Common food allergens include bananas, beef products, caffeine, chocolate, citrus fruits, corn, dairy products, eggs, oats, oysters, peanuts, processed foods, salmon, strawberries, tomatoes, wheat and white rice. Many people are sensitive to food additives, the most common being the dyes F, D and C yellow #5. Other common irritants are vanillin, benzyldehyde, eucalyptol, MSG, BHT, BHA, benzoates, and annato. Some simple solutions to problems caused by food allergies are:

- Avoid the foods and additives listed above.
- For 30 days, omit foods you suspect may be causing allergies. Then reintroduce them one at a time to see if you have an allergic reaction. This way you can single out the offending foods.

Enzyme Deficiency

Enzymes are necessary for food to be broken down and utilized as energy but too often enzymes are destroyed by cooking and other types of food processing. Raw fruits and vegetables are rich in enzymes, yet few Americans eat them regularly. The average American diet is enzyme-deficient which leads to bloating, stomach pain, gas, and low energy. Some solutions to these problems are:

- Eat at least four servings of raw fruits and five servings of raw vegetables daily.
- Take enzyme supplements with meals.

Herbs for Healthy Digestion

Overeating and Constant Snacking

Feasting at the Thanksgiving table is one thing; but eating every meal like it's your last is another. Regularly overeating puts a strain on the digestive system and can cause bloating, gas, and other forms of stomach/intestinal discomfort. Snacking also taxes the digestive system and rare is the American who isn't guilty of this pastime! From potato chips to cookies to soda to ice cream, we're surrounded by easy-to-grab snack foods and plenty of leisure time to indulge. Constant snacking makes the digestive system work overtime, and is a common cause of digestive discomfort. In addition to being unhealthy for the digestive system, both of these habits contribute to obesity. Two solutions are:

- Eat in moderation (although it's easier said than done).
- Avoid constant snacking. If you must snack, eat enzyme-rich raw fruits and vegetables.

Imbalance of Beneficial Bacteria

Antibiotics. Are they a blessing or a curse? When they were first discovered, they appeared to be a miracle drug. However, with time it has become clear that antibiotics are a two-edged sword. The downside is that they are creating more resistant strains of bacteria by killing off weaker bacteria, and leaving the strong to survive and multiply. And although antibiotics are effective in killing harmful organisms, they also kill off the "good guys" that keep the "bad guys" in check. For instance, penicillin destroys germs that cause illness, but it also kills the bacteria that prevents the growth of *Candida albicans,* a fungus that causes yeast infections.

When the colon's healthy bacteria population, including *Lactobacillus acidophilus,* is diminished, problems are inevitable. For starters, since these friendly flora help digest food, when they are in short supply stagnation, putrefaction and toxic buildup occur in the colon. Also, susceptibility to illnesses caused by unfriendly bacteria and yeasts increases. Allergies can be a symptom that the colon's

Herbs for Healthy Digestion

healthy bacteria population is suffering. Fortunately, the "good guys" can be reintroduced to the colon via supplements and healthy eating. Some solutions to the problems caused by bacteria imbalance are:

- Take supplemental *Lactobacillus acidophilus* along with other healthy bacteria cultures. It's available in liquid or capsule form.
- Eat a daily serving of yogurt containing active bacteria cultures. Brands that contain live cultures are labeled on the container.
- Avoid using antibiotics, if you can. Garlic, echinacea and goldenseal are excellent herbal antibiotics, and they don't harm the body's friendly bacteria. If you must use antibiotics, always take supplemental acidophilus along with them.

Poor Elimination

Poor elimination does not refer only to constipation (which is addressed in the next section). It also refers to not eliminating often enough, or not eliminating completely so that food is left to stagnate in the colon.

Processed grains (such as white flour and white rice), fast food, and prepackaged foods inhibit the normal elimination process. These foods lack fiber and create small, hard stools that have a long transit time in the colon and are difficult to eliminate. They also contain dozens of unhealthy ingredients like preservatives, dyes, and unhealthy fats. If food is allowed to sit in the colon for long periods of time it becomes a breeding ground for toxins that poison the whole body and cause disease. This phenomenon is called autointoxication, or self-poisoning.

Modern Western society is plagued with diet-related colon problems. More than 20 percent of Americans suffer from constipation. From irritable bowel to colitis to colon cancer, we endure problems unknown in countries where high-fiber diets are a way of life. It's no coincidence that laxatives are one of today's top-selling non-prescription drugs. People in Third World countries consume about two and a half times more fiber than those in Western countries. Standard transit time for food in the colon is 18 hours in undeveloped nations, whereas in the United States, food transit time can take from three days to two weeks.[2] Speaking of the dangers of slow elimination, Warren Levin, a holistic physician, says, "If the bowel content is small, it can take from 75 to 100 hours for the foods we eat to pass through. When this occurs, there is stagnation and putrefaction. Foods become toxic and we absorb toxins. This is one of the ways in which ill health is produced . . . ."[3]

A healthy colon eliminates remnants of undigested food, secretions from the intestines such as mucus and salt, and bacteria and parasites that are broken down from blood and tissues. In recent years it has been verified that healthy elimination even prevents diseases such as breast cancer, and that a fiber-rich diet lowers cholesterol and thereby helps prevent heart disease. Every function of the body is in some way influenced by the health of the colon. Some solutions to poor elimination are:

- Eat fiber-rich foods—fresh fruits, vegetables, whole wheat, brown rice, beans, lentils.
- Take a daily psyllium supplement.
- Avoid eating bread products made with white flour. The fiber in the whole grain has been processed out and such products are harder for the body to eliminate.

COMMON DIGESTIVE DISORDERS

CONSTIPATION

Constipation is the result of waste matter moving too slowly through the bowel. Chronic constipation leads to other problems like gas, insomnia, hemorrhoids, indigestion, diverticulitis, appendicitis, varicose veins, hernia, and colon cancer. The cause is usually lack of adequate dietary fiber, but some drugs cause constipation, as do some supplements (i.e., iron). It is also common during pregnancy.

Herbs for Healthy Digestion

Constipation is rampant in modern society. Americans spend more than $400 million a year on laxatives. Elimination should occur at least on a daily basis. Average transit time for food in the body is 18 to 24 hours. If food remains in the colon longer than that, toxins can form.

Constipation is usually easily corrected with wise food selection and use of herbs that promote elimination. Fiber, such as that contained in psyllium hulls, adds bulk and stimulates the muscles that promote elimination. Fiber is essential—not optional—for intestinal health. For mild constipation, additional fiber may be enough to correct the problem. If that doesn't work, try the herb cascara sagrada. For severe constipation, try strong herbs such as senna or rhubarb, which are powerful colon stimulants.

Helpful Herbs

Aloe vera juice—A powerful purgative (an herb that causes vigorous elimination). Purgative herbs sometimes cause intestinal griping, cramping and strong urgency to eliminate, but are useful for those who suffer from chronic constipation for whom milder remedies are ineffective. Because of its unpleasant taste, aloe vera juice is often taken in dried, capsulated form.

Cascara sagrada—The most widely used laxative in the world. This herb has no equal when it comes to curing constipation. It is an ingredient in many over-the-counter laxatives. Cascara is milder than the laxative herbs aloe, buckthorn, rhubarb, and senna. It is less likely to cause cramps or stomach discomfort, but in some people it may. If you use cascara, start with a small amount and if you tolerate it well, use more if necessary.

Flax seed—Flax seeds absorb liquid and swell in the intestines, encouraging elimination by increasing bowel contents. Drink plenty of water with them.

Licorice root—A mild laxative, especially appropriate for children.

Psyllium hulls—This valuable plant has been used since ancient times. The hulls absorb as much as 10 times their weight when

Herbs for Healthy Digestion

mixed with liquid and become mucilaginous (gel-like), adding bulk and lubrication in the intestines. Larger stools press on the intestines and stimulate the muscles that promote elimination. One teaspoon three times daily is usually sufficient to cure normal cases of constipation.

Senna—Because it is a powerful colon stimulant, senna should be a laxative herb of last resort. It can cause cramping and stomach pain. If you have constipation that doesn't respond to increasing fiber and water intake and exercising, and you've tried the gentler cascara sagrada, senna may in order. Start with small amounts and work to larger amounts if necessary.

Other Supplements

Apple pectin—Forms a gel when mixed with water and provides bulk and lubrication to the stool.

General Recommendations

- Get regular exercise, at least 30 minutes three times weekly. Exercise stimulates the elimination process.
- Eat plenty of fresh fruits and vegetables They are rich in fiber to promote healthy elimination.
- Eat prunes—their laxative effect is well-known and effective.
- Try linseed oil—it's a stool softener.
- Avoid dairy products, white flour and sugar. They tend to be constipating.
- Drink at least eight glasses of water a day. Water lubricates body tissues, cleanses, and promotes elimination.
- Avoid artificial laxatives. Regular use of them, over a year or more, may damage the colon and lead to dependency.
- If none of the suggestions or natural remedies offered here provide relief, try an enema as a last resort.

Herbs for Healthy Digestion

Diarrhea

Loose stools and a frequent urgency to eliminate are two well-known signs of diarrhea. Other common symptoms include thirst, abdominal pain, and sometimes vomiting and fever.

Diarrhea has many causes, including food poisoning, viruses and bacteria, contaminated water, lactose intolerance, sensitivity to the sweetener sorbitol, caffeine, or any other substance body may be sensitive or allergic to. Normal diarrhea is the body's way of cleansing out harmful substances and, unless it is chronic, should not be a cause for alarm. Most cases of diarrhea are self-limiting; that is, they run their course and clear up on their own.

Diarrhea that alternates with constipation can be a symptom of irritable bowel (discussed later in this section). Chronic diarrhea can be a symptom of Crohn's disease, colitis, diverticulosis, colon cancer and should be taken seriously. Many of the supplements suggested here can make a remarkable difference.

When food poisoning is the cause of diarrhea it usually hits within six hours of eating the offending food. The staphylococcus or clostridium bacteria are the most common culprits. In cases of salmonella or campylobacer bacteria poisoning, diarrhea usually occurs within 12 to 48 hours after eating. In normal cases of diarrhea, the main concern is replacing fluids and electrolytes lost in elimination.

Helpful Herbs

Kelp—Abundant in trace minerals to replenish those lost in elimination.

Garlic—Kills many bacteria and viruses, common causes of diarrhea.

Psyllium hulls—Adds bulk to the stool and absorbs water to return normalcy to bowel movements. Absorbs toxins.

Ginger root—Soothing for nausea; can be taken as a tea or in capsules.

Slippery elm bark—Forms a gel when mixed with water which soothes and nourishes the bowel.

Herbs for Healthy Digestion

Goldenseal—Contains the alkaloid berberine which clinical studies have shown to be effective against diarrhea related to infection with E. coli, dysentery, salmonella, giardia, and cholera.[4] Goldenseal can be used as a preventive against these bacteria, useful when planning trips to countries that may have poor water supplies.

Other Supplements

Charcoal—Absorbs toxins and gas.

Potassium, calcium, magnesium, zinc—During bouts of diarrhea the body loses these and other minerals, which need to be replenished.

Food enzymes—Help the body digest food; poorly digested food can cause diarrhea.

Acidophilus—Has antibacterial, antifungal and antiviral properties, produces vitamins and promotes healthy stools.

General Recommendations

- Eat plenty of high-fiber foods.
- Because diarrhea is the body's way of cleansing, avoid diarrhea medications at least for the first couple days after the onset of symptoms to give the body a chance to cleanse.
- Avoid milk products.
- Drink plenty of water to replace lost liquids and prevent dehydration. Also drink broths to replace nutrients (homemade broth from cabbage, potatoes, and carrots is a good choice).
- Avoid caffeine when you're suffering from diarrhea; it stimulates the intestine. It also acts as a diuretic, increasing urination and subsequent dehydration.
- Eat yogurt with active acidophilus cultures to increase beneficial bacteria in the colon.
- Mix 1/2 teaspoon honey and a pinch of salt in 8 ounces of fruit juice to replace fluids and electrolytes.
- Avoid the sweeteners mannitol and sorbitol.

Herbs for Healthy Digestion

Flatulence

Let's be honest—everyone passes gas. Some foods, like beans, produce gas in just about everyone because they contain starches which are difficult for the body to break down. For some people just about everything they eat causes gas, an anxiety-provoking and embarrassing problem.

Gas (flatulence) has many causes. It is created when bacteria present in the large intestine cause incompletely digested carbohydrates to ferment. Methane gas is responsible for the characteristic unpleasant odor. Lactose-intolerance is one cause of gas. Swallowing too much air by either chewing with the mouth open or eating too quickly can cause gas. A continual diet of junk foods which lack enzymes, do not digest or eliminate properly, and stagnate in the colon can also lead to chronic gas. Some people are sensitive to certain foods and experience bouts with gas whenever those foods are eaten.

In some people, healthy foods—including whole wheat, fruits and vegetables—cause gas. These foods provide indigestible carbohydrates (fiber) which feed the bacteria that live in the stomach and intestines. When the body grows accustomed to eating more fiber, the gas usually diminishes. Disease conditions that sometimes cause chronic gas include peptic ulcers, gallstones, irritable bowel syndrome, and food allergies.

Helpful Herbs

Ginger—Ginger has a soothing, antispasmodic effect on the intestinal system.

Fennel—Relaxes the smooth muscle lining of the digestive tract and helps expel gas. In some cultures fennel seed is chewed after meals.

Papaya—Contains the enzyme papain, which is similar to pepsin, and other enzymes which break down milk proteins and starches to help prevent gas.

Herbs for Healthy Digestion

Psyllium hulls—Provide bulk to move contents through the intestine, preventing the stagnation and resulting gas that can come from poor elimination. Use regularly.

Peppermint—As an antispasmodic, peppermint soothes the smooth muscle of the digestive tract.

Cinnamon—Helps break down fat in the digestive system, possibly by boosting enzyme action.

Other Supplements

Charcoal—Take an hour before meals. Charcoal effectively absorbs gas in the intestines.

Acidophilus—Populates the intestines with friendly bacteria which help digest food, keep bad bacteria in check and prevent gas.

General Recommendations

- Chew food thoroughly.
- Eat plenty of high-fiber foods (fresh fruits and vegetables, whole grains such as brown rice and whole wheat bread, legumes).
- If milk products give you flatulence, avoid them or only eat those especially for the lactose-intolerant. Yogurt is usually well-tolerated by those who are lactose-intolerant.
- Unfortunately, one of the most healthy high-fiber foods—beans—is also notorious for causing gas. Soak beans before cooking; then discard the soaking water and replace it with fresh water before boiling. Try the product Beano—it contains enzymes that help break down the indigestible fiber in beans.
- Take a walk after meals. Walking stimulates bowel activity, and helps push bowel contents along.
- Drink peppermint, chamomile, or fennel tea after meals.
- Cut back on dietary fat. It is hard to digest and can cause intestinal spasms.
- Avoid chewing gum; it increases the swallowing of air, creating painful gas.

Herbs for Healthy Digestion

Heartburn

Characterized by a painful burning in the throat just below the breastbone, heartburn occurs when stomach contents flow back into the esophagus. It is caused by too little hydrochloric acid (HCl) in the stomach, hiatal hernia, ulcers, spicy foods, fried foods, alcohol, coffee, citrus fruits, chocolate and a number of other foods.

If you regularly suffer from heartburn, take this simple test to determine whether supplemental hydrochloric acid would help you. When heartburn strikes, drink a tablespoon of lemon juice or apple cider vinegar. If the heartburn goes away, you need more HCl and can either take supplements with meals or sip apple cider vinegar in water with meals. If the heartburn gets worse, avoid taking supplemental HCl.

Although heartburn is rarely serious, it can be a symptom of a heart attack or ulcer. If you have the following symptoms along with heartburn, check with a health professional immediately as it could be a heart attack:

- Difficult or painful swallowing
- Vomiting with blood
- Bloody or black stool
- Shortness of breath
- Light-headedness, dizziness
- Pain in neck/shoulder

If heartburn worsens before meals, or if it occurs regularly over a period of three or four weeks, it could be an ulcer. Read the section on ulcers in this booklet for suggestions.

Helpful Herbs

Ginger—The most helpful of herbal remedies for heartburn. Take two capsules after you eat; increase the dosage as necessary.

Goldenseal—A bitter herb. Bitter herbs are traditionally used to aid digestion. Take in capsules or as a liquid extract just before eating.

Gentian—Another bitter herb. Take in capsules or as a liquid extract just before eating.

Herbs for Healthy Digestion

Fennel—Historically used as a heartburn treatment.
Catnip—Take in capsules or as a tea.

Other Supplements

Food enzymes (avoid those with hydrochloric acid)—Help the body break down and digest food.

General Recommendations

- Chew food thoroughly.
- Sit up after eating—lying down can allow stomach acid to sneak into the esophagus and cause heartburn pain.
- Avoid antacids that contain aluminum and/or sodium. Avoid prolonged use of antacids. They cover up the symptoms but don't get to the roots of the problem, and can cause mineral imbalance.
- Sip a teaspoon of apple cider vinegar in water along with meals, and avoid drinking other liquids while eating.
- Avoid carbonated drinks, sugar, and highly processed or fried foods.
- Don't smoke. Smoking increases susceptibility to heartburn by increasing stomach acid production.
- Lose weight. Pressure from extra weight can push stomach acid into the esophagus.
- Prescription antidepressants and sedatives may cause heartburn in some people.
- Stress may contribute to heartburn by increasing stomach acid production. Avoid eating when stressed.

ULCERS

An estimated 5 million Americans suffer from ulcers. Theories as to what cause them have varied through the years from stress, to spicy foods, to aspirin, to a pesky bacteria called *Helicobacter pylori*. The truth is, ulcers can be caused or aggravated by any or all of these things.

Herbs for Healthy Digestion

Ulcers are of two varieties—duodenal (in the small intestine near the stomach) or gastric (in the stomach lining). They are generally chronic; that is, they come and go.

Antiulcer diets of the past are outdated. For instance, milk used to be recommended to soothe ulcers, but now we know milk increases the production of stomach acid, which further aggravates ulcers. The belief that spicy foods irritate ulcers is also a thing of the past. Researchers who have filmed stomachs after spicy foods were eaten detected no signs of irritation.

Helpful Herbs

Aloe vera juice—Heals irritated tissues. Use prudently, as aloe vera juice is a powerful cathartic.

Capsicum—Studies have shown that capsicum does not irritate the stomach, as popular belief would have it. Capsicum is a longtime folk remedy for ulcers. One of this century's most beloved herbalists, the late Dr. John Christopher, was an enthusiastic supporter of capsicum for ulcer treatment.

Chamomile—Experimental animals given chamomile along with ulcer-causing chemicals developed significantly fewer ulcers than animals not fed chamomile. Those who already had ulcers recovered more quickly when fed chamomile than those who weren't.[5]

Evening primrose oil—The essential fatty acids contained in evening primrose oil are healing to mucus membranes.

Goldenseal—This herb has antibiotic properties, therefore helping destroy harmful bacteria in the stomach.

Licorice root—Studies published in the most prestigious medical journals, including *Lancet* and *The Journal of the American Medical Association*, have lauded the virtues of licorice root for the treatment of ulcers. Studies comparing the active ingredient in licorice (glycyrrhetinic acid or GA) to Tagamet showed both were equally effective against duodenal ulcers, while Tagamet was more effective against stomach ulcers. However, GA provides more protection against recurrence.[6] Because GA causes water reten-

Herbs for Healthy Digestion

tion, deglycyrrhizinated licorice (DGL) is widely available today.

Papaya—Appears to have a protective effect against ulcers. In one study, two groups of were animals fed aspirin and steroids to induce ulcers. One group was also fed papaya prior to being given the ulcer-inducing drugs, and developed significantly fewer ulcers. Papaya may be taken as a preventive measure.[7]

Slippery elm—Forms a gel-like substance when mixed with water, which is soothing and nourishing to mucus membranes.

Other Supplements

L-Glutamine—Necessary for healthy mucus production to protect stomach lining. The success of cabbage juice in the treatment of ulcers may be linked to its glutamine content.

Vitamins A and E—These vitamins have been shown to inhibit the development of stress-caused ulcers in rats.[8] They are also essential in nourishing and protecting the lining of the digestive tract.

Iron—Useful for those who suffer from a bleeding ulcer.

Zinc—Increases the production of mucin, a protective substance that is a main component of mucus. In studies, zinc demonstrated a protective effect against peptic ulcers.[9]

General Recommendations

- Drink up to two quarts daily of freshly made cabbage juice (made in a juicer). Studies show raw cabbage juice has amazing success in the healing of ulcers, possibly due to its glutamine content.[10] Glutamine is necessary for the production of protective mucus.
- Avoid using aspirin.
- Avoid milk products.
- Avoid caffeine, tea, alcohol, and carbonated beverages.
- Don't smoke; smokers are twice as likely to develop ulcers.
- Food allergies are a prime suspect in causing ulcers. Elimination diets have proven successful in treating many ulcer patients. If food allergy is the cause, the ulcer will not heal until the offending food is eliminated. For 30 days, omit foods you suspect may

be causing allergies. Then reintroduce them one at a time to see if you have an allergic reaction. This way you can single out the offending foods.
- Eat plenty of fiber; a fiber-rich diet is associated with reduced risk of duodenal ulcers.

IRRITABLE BOWEL SYNDROME (IBS)

This disorder, which afflicts as many as 15 percent of all people at some time in their life, is second only to the cold in frequency of occurrence. It is the most common disorder of the gastrointestinal system.

Irritable bowel syndrome is also commonly known as spastic colitis, mucus colitis, and nervous indigestion. Symptoms include pain in the abdomen, mucus in the stool, alternating diarrhea with constipation, nausea, flatulence, and bloating. Because symptoms mimic other disorders, it is important to consult your health care professional to rule out any of the following ailments:

- Allergy or sensitivity to certain foods or beverages
- Bacterial infection (such as giardia)
- Crohn's disease or ulcerative colitis
- Candida infection in the intestines
- Decreased friendly bacteria population due to use of antibiotics
- Diverticulitis
- Cancer

IBS can be caused by a variety of physical, emotional and dietary factors. Food allergies are a common cause, as is stress. Lack of adequate dietary fiber is also a culprit. Treating IBS naturally falls into four simple steps: increase fiber, eliminate allergenic foods, control stress, and build colon health with supplements.

Herbs for Healthy Digestion

Helpful Herbs

Garlic—Kills a wide variety of microorganisms and builds a healthful environment for friendly bacteria in the colon.

Ginger—Has been used for thousands of years to relieve the pain of intestinal gas. Also effective for relief of nausea. Studies have shown ginger to be more effective than the drug Dramamine in relieving motion-sickness.[11]

Peppermint oil—Inhibits contractions of the intestinal system and relieves gas. Studies have shown that enteric coated peppermint oil (specially designed to be released in the colon) significantly improved abdominal distress of IBS sufferers.[12]

Psyllium hulls—Help regulate elimination by providing bulk and lubrication for healthy bowel movements. Well-tolerated, even by those who may be allergic to other fiber products such as wheat bran. A remedy for both diarrhea and constipation, which often plague IBS sufferers.

Slippery elm—When mixed with water, slippery elm forms a gelatinous substance that is soothing to mucus membranes of the intestinal tract, and also provides nourishment.

Caprylic acid—Kills *Candida albicans* yeast, believed to be a cause of IBS. Candida is normally kept in check by digestive secretions and friendly flora, but use of antibiotics can cause it to proliferate.

Chamomile—Has a demonstrated antispasmodic effect on the intestinal system.[13]

Other Supplements

Food enzymes—Help the body digest and assimilate food; take stress off the digestive system.

Acidophilus—Friendly bacteria that inhabit the colon and keep unfriendly bacteria in check.

Essential fatty acids—Soothing and healing to mucus membranes of the gastrointestinal tract.

Herbs for Healthy Digestion

General Recommendations

- Foods to avoid: dairy products, animal fats, spicy foods, fried foods, wheat products, all junk foods, carbonated beverages, alcohol and smoking.
- Try an elimination diet: Eat only hypoallergenic foods such as bananas, chicken, potatoes, rice, and apples for a week. If IBS symptoms disappear, food allergy is the likely cause of suffering. After a week, introduce one commonly eaten food every two days and observe whether IBS symptoms begin. Continue introducing foods one at a time, keeping a record of how the body reacts.
- Drink at least six to eight glasses of water a day.
- Stay away from caffeine.
- Avoid products with the sweetener sorbitol.
- Get regular, moderate exercise.
- To calm spasms, use a hot water bottle or heating pad on stomach.

DIVERTICULOSIS

Diverticulosis occurs when mucus membranes in the large intestine become inflamed and form small, pouchlike areas, called diverticula. Diverticulosis is rare in those younger than 35, but as many as one in ten Americans over 40 has them, and half of those 60 and older. It's rare in Third World countries, and is largely attributable to Western lifestyle.

Usually, diverticulosis has no symptoms. However, in some people—about 10 percent—the diverticula become blocked off, infected with bacteria and inflamed, a condition called diverticulitis. As with other colon disorders, lack of fiber is a prime cause. Other risk factors include poor eating habits, a family history of the disease, obesity, gall bladder disease and coronary artery disease.

Symptoms of diverticulitis include cramping, tenderness on the left side of the abdomen relieved by passing gas or having a bowel movement, constipation, diarrhea, and nausea.

Herbs for Healthy Digestion

Helpful Herbs

Psyllium hulls—Eating plenty of fiber is the best way to treat and prevent diverticulitis. Mixed with liquid, psyllium hulls swell up several times their original size and speed up the passage of food through the colon, decreasing risk of fecal matter becoming impacted and infected.

Slippery elm—Nourishing to tissues of the colon, mild and easily digested.

Garlic—Kills many types of bacteria, a cause of inflammation in diverticulitis.

Other Supplements

Charcoal—Absorbs toxins and gas in the colon.

Food enzymes—Assist with digestion and assimilation of food.

Acidophilus—Populate the colon with friendly bacteria, which help digest food, produce nutrients and keep harmful bacteria in check.

Free form amino acids—Easily digested form of protein to provide essential nutrition needed for healing.

Chlorophyll—Helpful for cleansing and healing irritated tissue.

General Recommendations

- Eat whole grains rather than processed ones (brown rice, whole wheat).
- Blend fruits and vegetables in a blender before eating. Eat steamed vegetables. As tolerance increases add raw fruits and vegetables. Drink vegetable juices such as carrot, cabbage.
- Use enemas to clean trapped fecal matter and relieve discomfort.
- Massage left side of abdomen to relieve pain.
- Drink six to eight glasses of water daily.
- Exercise regularly—exercise tones the colon.
- Heed the call of nature—when you need to eliminate, do so.

Herbs for Healthy Digestion

HEMORRHOIDS

Hemorrhoids are a common ailment, affecting as many as eight out of ten people at some time in their lives. Susceptibility to them can be hereditary, but the main cause is straining when having a bowel movement due to lack of dietary fiber. Such straining weakens veins around the anus, causing protrusions to form. They may itch, bleed and tear, causing great pain. The simplest way to treat hemorrhoids is with increased dietary fiber. Lack of exercise, heavy lifting, prolonged sitting, obesity and pregnancy can also increase the risk of hemorrhoids.

Helpful Herbs

Psyllium hulls—By softening and increasing the bulk of the stool, psyllium hulls provide significant relief from hemorrhoids. Used daily, they can also prevent hemorrhoids.

Guar gum—Like psyllium, guar gum is a mucilaginous, soothing fiber product and also acts as a mild laxative.

Butcher's broom—An astringent herb which helps shrink inflamed tissues.

Aloe vera juice—Apply directly to external hemorrhoids for soothing relief.

Blackberry—A traditional herbal remedy, blackberry has astringent action.

Mullein—Studies have shown that mullein has anti-inflammatory and astringent properties due to its tannin content.[14]

Other Supplements

Vitamin C with bioflavonoids—Strengthens capillaries, helpful for healing and prevention of hemorrhoids.

Pycnogenol—Pycnogenol products on the market are derived from pine bark and/or grape seed. These are powerful antioxidants and they provide nutrients that help strengthen vein structures.

Herbs for Healthy Digestion

General Recommendations

- Eat a high fiber diet—plenty of fresh fruits, vegetables and whole grains.
- Get regular exercise.
- Drink plenty of liquids.
- Avoid colored and scented toilet paper. Dampen toilet paper before wiping. Substitute facial tissues with aloe or moisturizing cream for regular toilet paper.
- Avoid heavy lifting.
- Take a sitz bath for immediate relief—sit with your knees raised in three or four inches of water in a bathtub. Warm water increases circulation to the area and provides relief.
- Apply witch hazel with a cotton ball to the irritated area.
- Avoid excess salt; it can make irritation worse.
- Avoid caffeine; it can cause rectal itching.

CONCLUSION

The intestinal problems that plague modern Western society are largely a product of poor diet. The single most important way to prevent such problems is to eat plenty of fiber in the form of whole grains and fresh fruits and vegetables. In addition, Mother Nature offers countless natural products that cure and prevent digestive problems. By following the dietary recommendations in this booklet, most people can enjoy a lifetime of healthy digestion.

Endnotes

1. *American Journal of Public Health*, 80:1443, 1990.
2. Morton Walker, D.P.M. "Internal Cleansers—Part I," *Health Foods Business*, September 1989, 58.
3. Morton Walker, D.P.M., "Internal Cleansers—Part II," *Health Foods Business*, September 1989, 29.
4. Michael Murray and Joseph Pizzorno, *Encyclopedia of Natural Medicine*, (Rocklin, CA: Prima Publishing, 1991), 289.
5. Michael Castleman. *The Healing Herbs* (Rodale Press, 1991), 109.
6. Ibid., 237-238.
7. Ibid., 278.
8. P.L. Harris, E.L. Hove, M. Mellott and K. Hickman, "Dietary production of gastric ulcers in rats and prevention by tocopherol administration," *Proc. Soc. Exp. Biol. Med.,* 1947, 4, 273-277.
9. D.J. Fommer, "The healing of gastric ulcers by zinc sulphate," *Med. J. Austr.,* 1975, 2, 793.
10. G. Cheney, "Rapid healing of peptic ulcers in patients receiving fresh cabbage juice," *Ca. Med.,* 1949, 70, 10-14.
11. D. Mowrey and D. Clayson, "Motion sickness, ginger, and psychophysics," *Lancet,* 1982, i, 655-657.
12. W. Rees, B. Evans and J. Rhodes, "Treating irritable bowel syndrome with peppermint oil," *British Medical Journal,* 1979, ii, 835-836.
13. H.B. Foster, H. Niklas and S. Lutz, "Antispasmodic effect of some medicinal plants," *Planta Medica,* 1980, 40, 309-319.
14. Michael Castleman, *The Healing Herbs,* (Rodale Press, 1991) 266.

Recommended Reading

Balch, James F. and Phyllis. *A. Prescription for Nutritional Healing.* Garden City Park, NY: Avery Publishing, 1990.

Castleman, Michael. *The Healing Herbs.* Emmaus, PA: Rodale Press, 1991.

Deborah Tkac, Ed. *The Doctor's Book of Home Remedies.* Rodale Press, 1991.

Lust, John. *The Herb Book.* New York, NY: Bantam Books, 1974.

Melville, Arabella, and Johnson, Colin. *Health Without Drugs.* New York, NY: Simon and Schuster, 1987.

Murray, Michael, N.D. and Pizzorno, Joseph, N.D. *Encyclopedia of Natural Medicine.* Rocklin, CA: Prima Publishing, 1991.

University of California at Berkeley. *The New Wellness Encyclopedia.* New York, NY: Houghton Mifflon Co., 1995.

Van Amerongen, C. Book of the Body: *The Way Things Work.* New York, NY: Simon and Schuster, 1979.